PE PREPARED

CIVIL PE PRACTICE EXAM

BREADTH EXAM

VERSION C

Congratulations on your decision to take the Principles and Practice of Engineering Exam for Civil Engineering!

PE Prepared was created by real, practicing civil engineers to give E.I.T.s and E.I.s like yourself a leg up on test day. We strove to author realistic questions at the right level of difficulty, with detailed, step-by-step solutions to help you learn the content that is going to be on the exam.

Our questions aren't harder than they need to be, but aren't easier either. They should take less than 6 minutes to solve. Take PE Prepared practice exams as a realistic simulation of exam day to measure your level of preparedness, or simply use them as a bank of practice questions while you study. The choice is yours!

Remember: Civil engineering is a noble profession. Civil engineers make the world a better, safer, and healthier place for people to live in. Congratulations again on your decision to take the PE exam, you're going to pass!

Check out our website at **PEprepared.com** for video workshops, study tips, and other resources for the exam. We can be contacted at ask@PEprepared.com.

START

CIVIL PE PRACTICE EXAM

BREADTH EXAM

VERSION C

101. Approximately 300 feet of new water main will be installed per the trench detail below. With a 10% allowance for waste, how many tons of pipe bedding material should be ordered? Assume the pipe bedding density is 135 pounds per cubic foot.

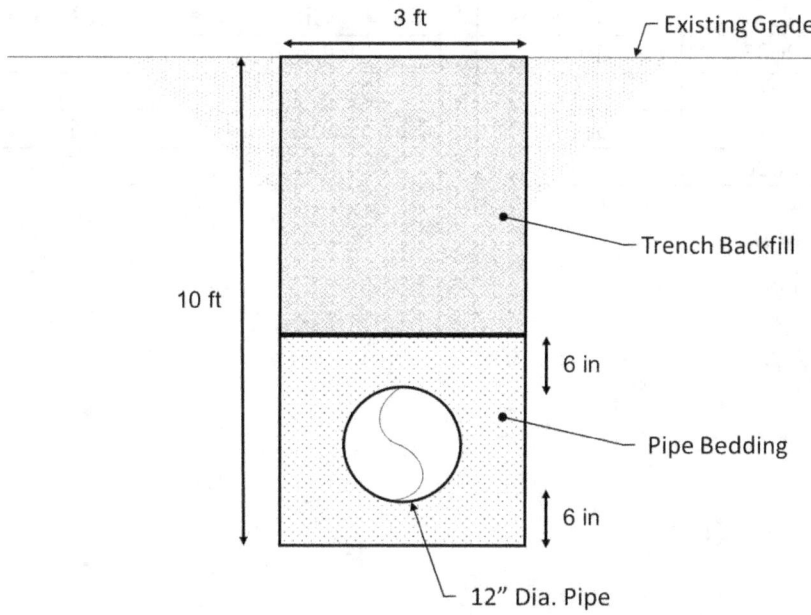

(A) 106 tons
(B) 116 tons
(C) 122 tons
(D) 535 tons

102. A project was completed in Dallas, TX in 2006 for $1,200,000. An equal project is planned for 2018 in San Francisco, CA. Using the information below and linear interpolation, the approximate cost for the new project is most nearly:

Year	Cost Index
2005	1.00
2010	1.13
2015	1.27
2020	1.47

City	Cost Index
Dallas, TX	0.98
Denver, CO	1.00
San Francisco, CA	1.18
New York, NY	1.20

(A) $1,444,900
(B) $1,625,800
(C) $1,957,500
(D) $1,968,200

103. An excavator and operator can excavate 10 cubic yards per hour over a ten-hour workday. The hourly excavator cost is $100 and the straight time hourly operator cost is $50. Overtime is paid for all hours over eight at a rate of 1.5 times straight time. Assume ten-hour work days. The amount budgeted to excavate 165 cubic yards should most nearly be:

(A) $950
(B) $2,525
(C) $2,550
(D) $2,575

104. A CPM schedule is shown below. Which of the following activities cannot be delayed without impacting the total duration of the project?

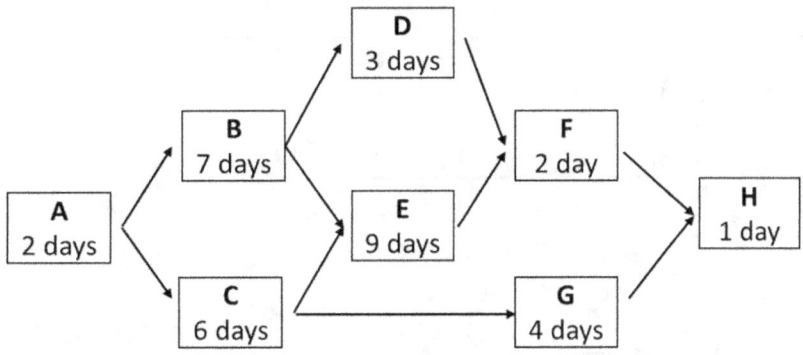

(A) Activity G
(B) Activity B
(C) Activity C
(D) Activity D

105. Examine the figure below, with a load centered on an ideal pulley. Assume rigging must be sized to support 5 times the maximum load applied. The minimum rigging breaking strength (ton) to use is most nearly:

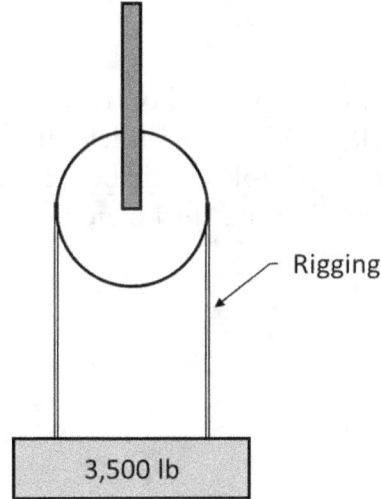

(A) 1 ton
(B) 2 ton
(C) 5 ton
(D) 9 ton

106. The figure below is not to scale. What is most nearly the minimum required slope for the sanitary sewer pipe to achieve 18" separation from the domestic water when the two utilities cross? Neglect pipe wall thicknesses.

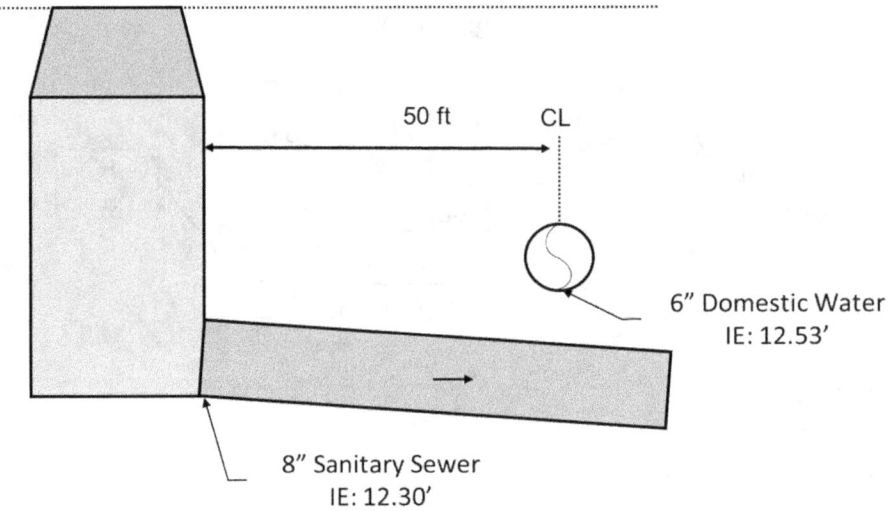

(A) 0.5 %
(B) 2.5 %
(C) 3.2 %
(D) 3.9 %

107. A steel plate is placed across an open utility trench overnight so vehicular traffic can cross. The trench is 10' wide. The largest vehicle that will cross weighs 5,000 pounds. Assume the steel plate acts as a simple elastic beam with hinge supports. What is most nearly the maximum moment that the steel plate will experience?

(A) 5 kip ft
(B) 13 kip ft
(C) 25 kip ft
(D) 63 kip ft

108. The difference in magnitude between the active earth pressure and passive earth pressure on the wall below is most nearly:

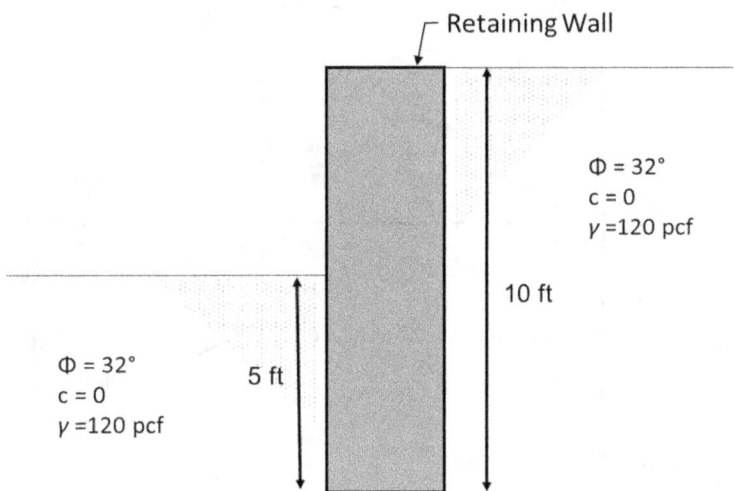

(A) 1,860 lb
(B) 3,015 lb
(C) 4,875 lb
(D) 19,035 lb

109. Which of the following statements regarding soil consolidation is the most correct?

I. An overconsolidated clay's consolidation curve is steeper when overburden pressure exceeds the preconsolidation pressure.
II. A normally consolidated clay's consolidation curve is steeper when overburden pressure exceeds the preconsolidation pressure.
III. An overconsolidated clay's consolidation curve is less steep when overburden pressure exceeds the preconsolidation pressure.
IV. A normally consolidated clay's consolidation curve is less steep when overburden pressure exceeds the preconsolidation pressure.

(A) I
(B) II
(C) III
(D) IV

110. What is most nearly the effective stress at the bottom of the clay layer shown below? Assume γ_d = dry unit weight, γ_b = buoyant unit weight, and w = water content.

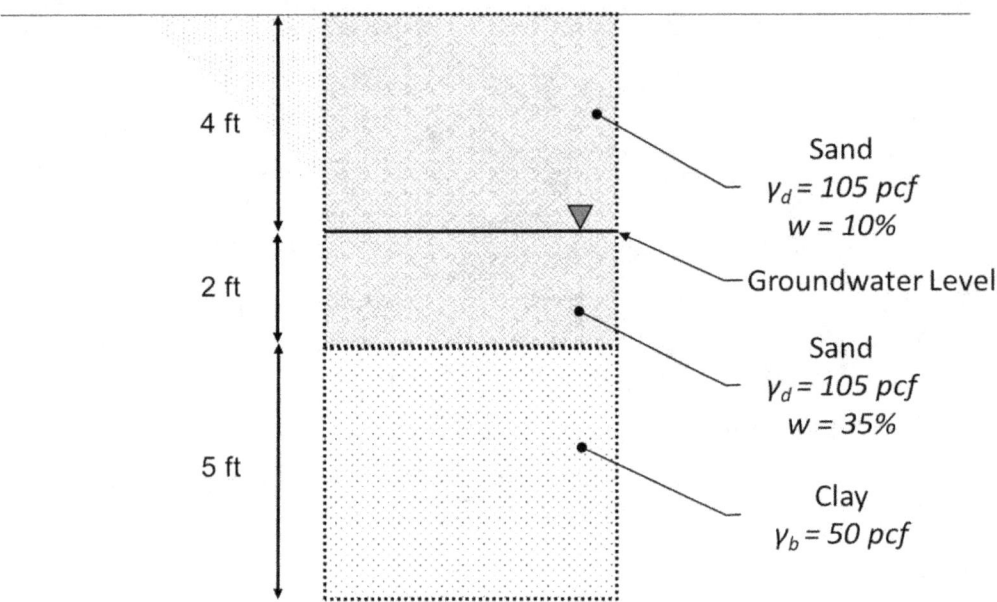

(A) 443 psf
(B) 871 psf
(C) 880 psf
(D) 996 psf

111. Using a factor of safety of 3, the allowable bearing capacity for the strip footing shown below is most nearly:

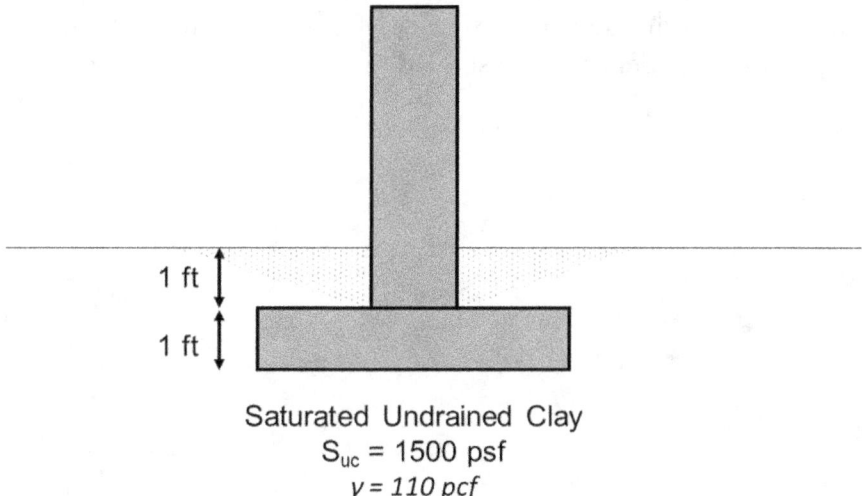

1 ft

1 ft

Saturated Undrained Clay
S_{uc} = 1500 psf
$\gamma = 110\ pcf$

S_{uc} = unconfined compressive strength.

(A) 1,360 psf
(B) 1,580 psf
(C) 2,570 psf
(D) 3,855 psf

112. Use the information given below regarding consolidation of a clay under loading from a newly constructed foundation. The clay layer is 20-feet thick and drainage is through the top and bottom surfaces. How long will it take to reach 90% of the long-term primary consolidation?

Coefficient of Consolidation c_v = 0.0029 square feet per day

(A) 8 years
(B) 25 years
(C) 80 years
(D) 150 years

113. A 10-foot high fill slope is constructed at a 2:1 (horizontal to vertical) slope in a saturated clay with cohesion equal to 500 psf and a total unit weight of 120 pounds per cubic foot. The clay is underlain by a firm base at a depth of 5 feet. Using a slope stability chart for undrained, cohesive soils ($\Phi=0$), determine the factor of safety against slope failure under worst-case loading.

(A) 1.5
(B) 2.6
(C) 5.5
(D) 6.3

114. Assume a dead load, D, of 1000 psf, a live load, L, of 750 psf, a snow load S, of 250 psf, a wind load, W, of 300 psf, and an earthquake load, E, of 500 psf. Which of the load combinations below results in the most conservative design?

I.   1.4D
II. 1.2D+1.6L+.5S
III. 1.2D+W+L+.5S
IV. 1.2D+E+L+0.2S

(A) I
(B) II
(C) IV
(D) V

115. The magnitude of the axial load in Member "A" is most nearly:

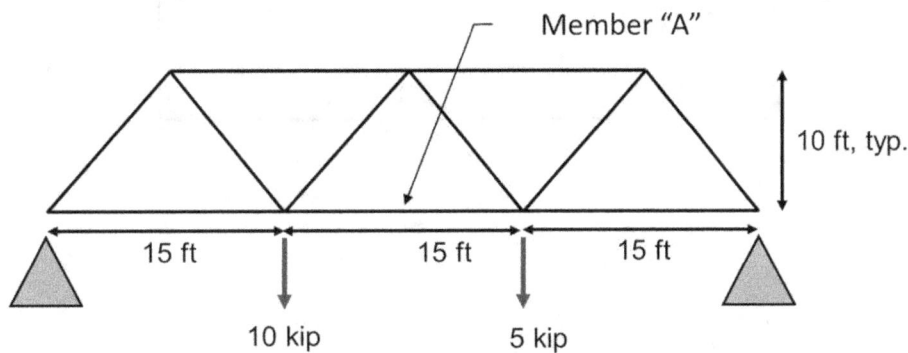

(A) 0 kip
(B) 6.7 kip
(C) 8.3 kip
(D) 11.2 kip

116. A beam is shown below. The magnitude of the bending stress at Point "A" is most nearly:

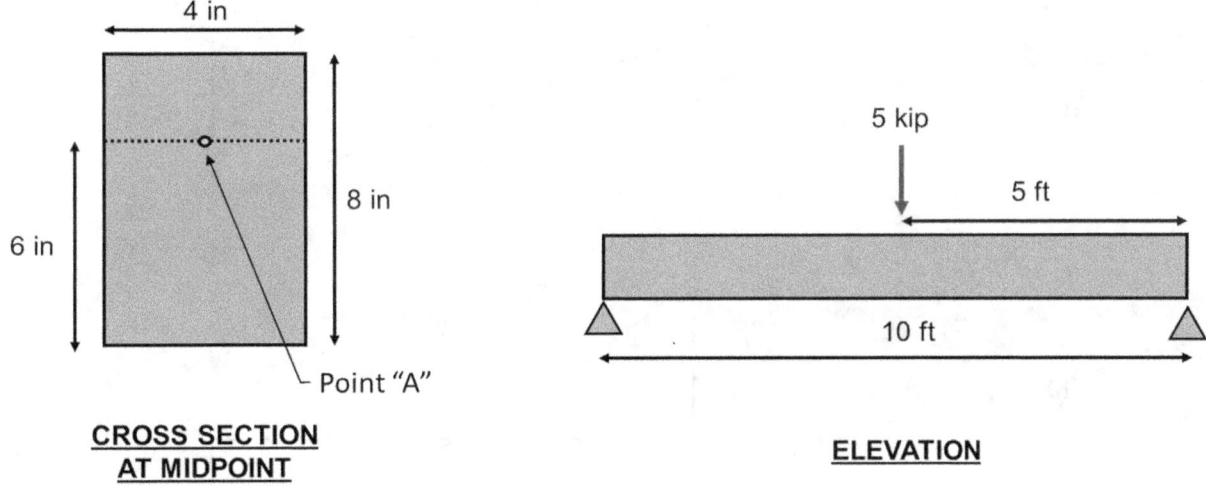

**CROSS SECTION
AT MIDPOINT**

ELEVATION

(A) 1.8 ksi
(B) 3.5 ksi
(C) 6.2 ksi
(D) 7.0 ksi

117. Select the proper shear diagram for the beam shown below:

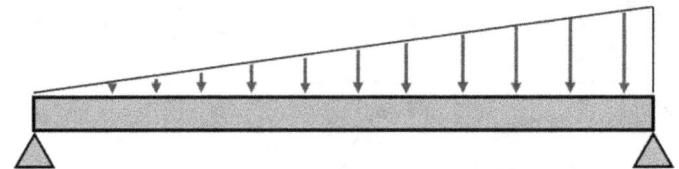

(A)

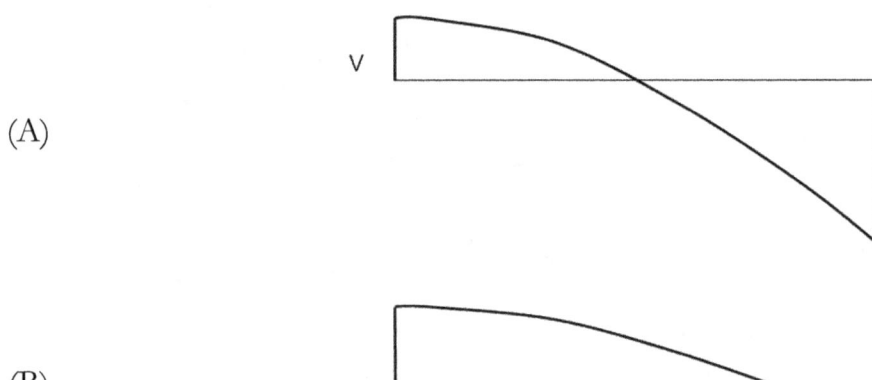

(B)

(C)

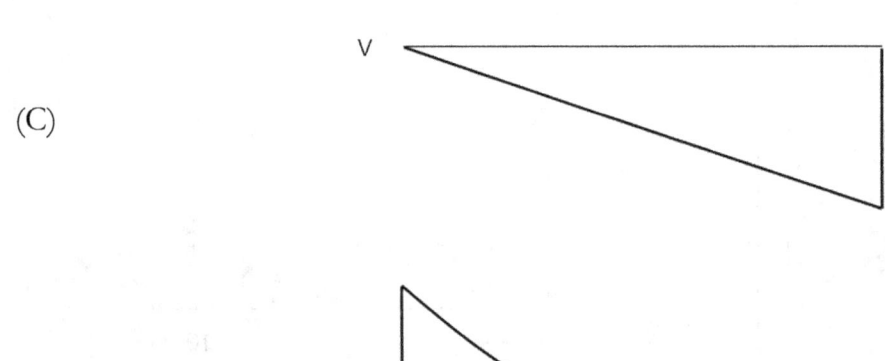

(D)

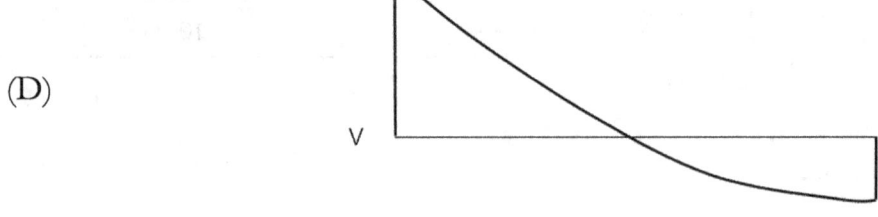

118. Examine the beam below. Information is provided for various beam shapes. Which is the most economical beam shape that should be selected to limit deflection to less than 0.14 inch?

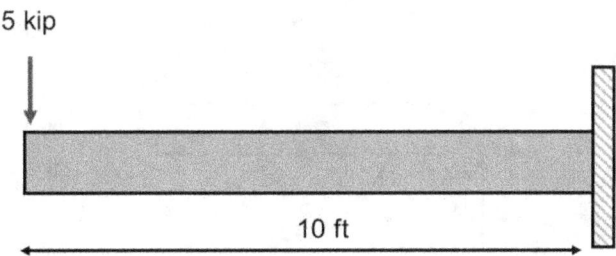

5 kip

10 ft

Shape	Z_x (in³)	I_x (in⁴)	E (10⁶ psi)
W14x86	126	1140	29
W14x74	126	795	29
W12x79	119	662	29
W10x88	113	534	29

(A) W14x86
(B) W14x74
(C) W12x79
(D) W10x88

119. A 100-foot long, 2-inch diameter steel cable has a modulus of elasticity of 30,000 ksi. How much tension is required for the cable to increase in length by 0.5 inch?

(A) 0.3 kip
(B) 4 kip
(C) 40 kip
(D) 471 kip

120. The triangular channel shown below is flowing at a constant depth, at a slope of 1%. Assume a Manning roughness coefficient of 0.02. If the channel discharges to a detention pond with vertical sidewalls and a bottom area of 1,000 square feet, how long will it take for the pond to reach a depth of 3 feet?

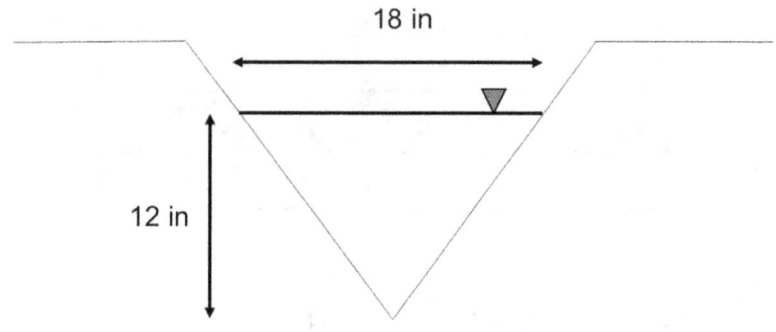

(A) 1.6 min
(B) 15 min
(C) 20 min
(D) 35 min

121. The peak runoff discharge rate from a parking lot is 5 cubic feet per second. The onsite stormwater management system is capable of treating 1,000 gallons per minute, and the remaining flow is bypassed through a PVC (n=0.013) overflow pipe sloped at 2%. The pipe must be designed to convey the design flowrate without surcharging. The minimum required diameter for the overflow pipe is most nearly:

(A) 8 in
(B) 10 in
(C) 12 in
(D) 18 in

122. Cumulative rainfall distributions for the SCS Type 1A and SCS Type 2 dimensionless unit hydrographs are shown below.

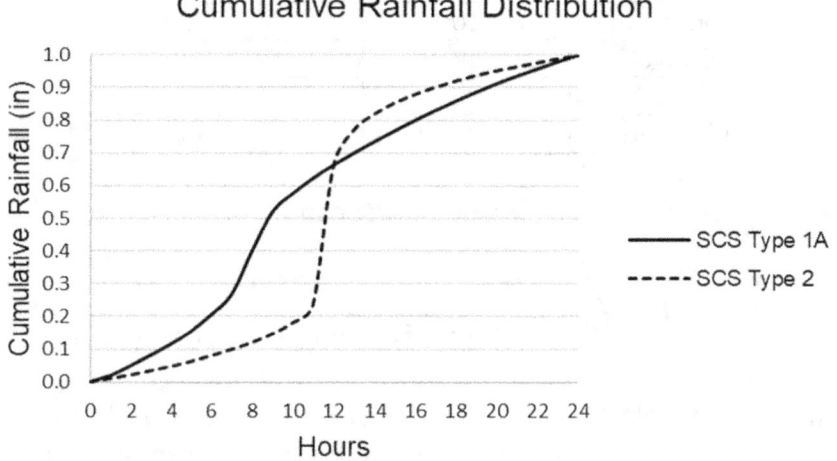

Which of the following statements is the most correct?

(A) The SCS Type 2 storm has a greater peak discharge than the SCS Type 1A storm.
(B) The SCS Type 1A storm has greater total precipitation than the SCS Type 2 storm.
(C) The SCS Type 1A storm will result in more runoff volume than the SCS Type 2 storm for any given basin.
(D) The peak discharge rate for the SCS Type 1A occurs later than the peak discharge rate for the SCS Type 2 storm.

123. A 25-acre drainage basin has the following land cover:

Land Cover	Area (ac)	Rational Method Runoff Coefficient
Landscaping	10	0.40
Gravel	4	0.75
Asphalt	6	0.85
Concrete	5	0.90

The rainfall intensity for the 100-year storm can be expressed as:

$$I(^{in}/_{hr}) = \frac{200}{time\ of\ concentration\ (min)}$$

If the time of concentration for the 100-year storm is one hour, the the maximum runoff rate for the 100-year storm using the Rational Formula is most nearly:

(A) 55 cfs
(B) 75 cfs
(C) 120 cfs
(D) 3,320 cfs

124. An empty 50-foot long by 50-foot wide rectangular stormwater pond with vertical sidewalls infiltrates water at a constant rate of 10 inch per hour. A 12-inch diameter outfall begins discharging to the pond at a rate of 500 gallons per minute. How high will the water level in the pond be in one hour?

(A) 1 in
(B) 8 in
(C) 9 in
(D) 24 in

125. A pipe is 8 inches in diameter and conveys 3,000 gallons per minute over a length of 100 feet. Assuming a Darcy friction factor of 0.018 and a Hazen-Williams coefficient of 120, which statement below is the most correct?

    (A) The Darcy-Weisbach equation predicts more headloss than the Hazen-Williams equation.
    (B) The Hazen-Williams equation predicts more headloss than the Darcy-Weisbach equation.
    (C) The headlosses predicted by the Darcy-Weisbach equation and the Hazen Williams equation are equal.
    (D) There is not enough information to calculate the headloss.

126. The energy grade line and hydraulic grade line are shown below for a certain piping system. Which dimension represents the total headloss through the system?

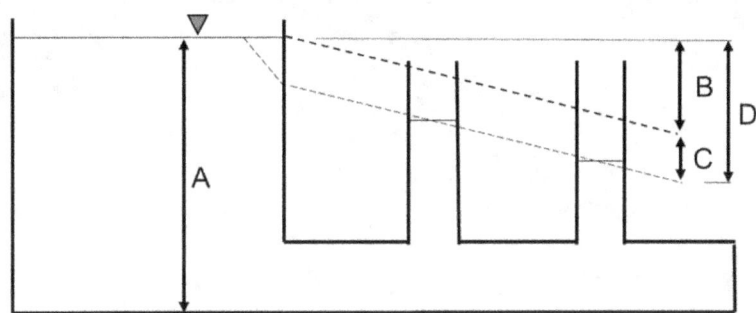

    (A) A
    (B) B
    (C) C
    (D) D

127. A horizontal curve has a long chord distance between the point of curvature and the point of tangency of 850 ft. The intersection angle is 25 degrees. The degree of curvature is most nearly:

(A) 2.9 °
(B) 5.4 °
(C) 6.8 °
(D) 8.2 °

128. A crest vertical curve transitions from a 2.5% grade uphill to a 0.5% grade downhill. Of the curve lengths below, which should be selected to achieve a minimum K-value of 120?

(A) 100 feet
(B) 200 feet
(C) 300 feet
(D) 400 feet

129. The following data was collected during a traffic count in 10-minute intervals. The peak hour factor is most nearly:

Time Interval	Count
5:00-5:10	150
5:10-5:20	160
5:20-5:30	145
5:30-5:40	155
5:40-5:50	162
5:50-6:00	148
6:00-6:10	145
6:10-6:20	138
6:20-6:30	148
6:30-6:40	152
6:40-6:50	140
6:50-7:00	135

(A) 0.95
(B) 1.03
(C) 1.06
(D) 5.67

130. Consider the Unified Soil Classification System. The most essential criteria for distinguishing between low plasticity clays (CL) and high plasticity clays (CH) is:

(A) Plastic Limit
(B) Plasticity Index
(C) Water Content
(D) Liquid Limit

131. The California Bearing Ratio (CBR) is a means of estimating which of the following?

(A) A soil's bearing capacity factors for general shear
(B) A soil's standard penetration resistance
(C) A soil's angle of internal friction
(D) A soil's suitability as a roadway subgrade

132. Concrete will be placed directly against extremely alkaline soils. The concrete mix would most likely consist of:

(A) Coarse aggregate, sand, Type II cement
(B) Coarse aggregate, sand, Type V cement, water
(C) Coarse aggregate, sand, Type I cement, water
(D) Coarse aggregate, Type III cement, water

133. The stress-strain curve below was obtained from tensile tests on a steel sample. The yield stress of the sample is most nearly:

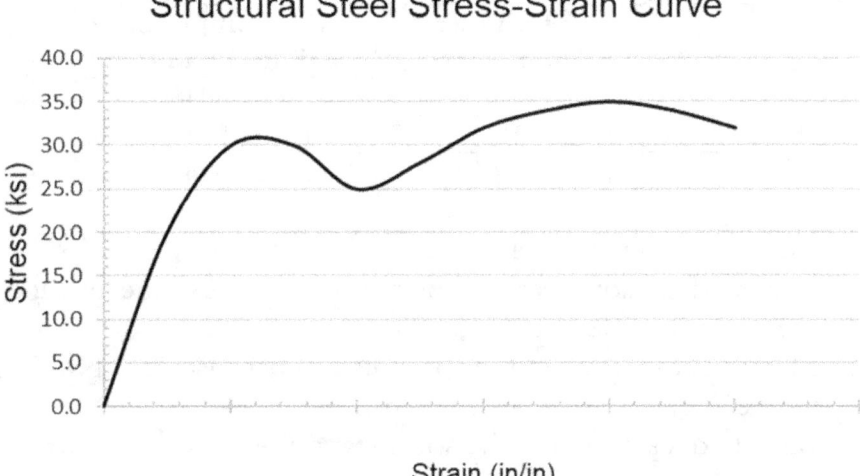

(A) 20 ksi
(B) 25 ksi
(C) 31 ksi
(D) 35 ksi

134. A soil sample is found to have a maximum dry density, as determined by the modified proctor, of 135 pounds per cubic foot and an optimum moisture content of 8%. If the soil must be compacted to 95% of the maximum dry density determined by the modified proctor and be at optimum moisture content, the minimum required wet density will most nearly be:

(A) 119 pcf
(B) 128 pcf
(C) 135 pcf
(D) 139 pcf

135. A specification requires the average of all concrete compressive strength test results to be greater than 3,000 psi with no single result less than 2,500 psi. Examine the results below and select the answer that is the most correct.

Test No.	Cylinder Diameter (in)	Axial Compressive Failure Load (lb)
1	6	91,584
2	6	92,574
3	6	70,548

(A) The concrete meets the specification.

(B) The concrete does not meet the specification because the average compressive strength of all tests is less than 3,000 psi.

(C) The concrete does not meet the specification because Test No. 2 resulted in a compressive strength less than 2,500 psi.

(D) The concrete does not meet the specification because Test No. 3 resulted in a compressive strength less than 2,500 psi.

136. A soil has a swell factor of 1.25 and a shrinkage factor of 0.15. If you excavate 100 compacted cubic yards, the number of loose cubic yards you'll need to haul away is most nearly:

(A) 115 LCY

(B) 118 LCY

(C) 125 LCY

(D) 147 LCY

137. The stakes shown below are encountered on a construction site at the locations shown. The finished grade elevation difference between Point 1 and Point 2 is most nearly:

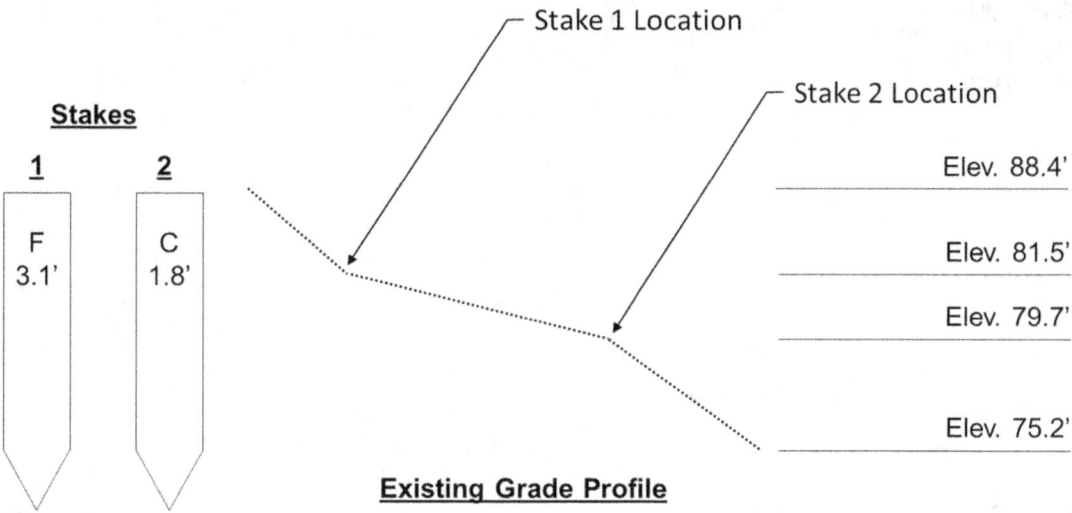

(A) 1.3 ft
(B) 1.8 ft
(C) 4.9 ft
(D) 6.7 ft

138. Temporary construction sediment and erosion control provisions are primarily intended to:

(A) Protect water quality and prevent streambank erosion.
(B) Prevent slope failures during rain events.
(C) Prevent property damage caused by construction.
(D) Minimize the development of impervious surfaces.

139. The safety regulations that apply specifically to the construction industry are located in which section of the Code of Federal Regulations?

   (A) 1903
   (B) 1904
   (C) 1910
   (D) 1926

140. As defined by OSHA, a "competent person" for the purpose of determining the safety of soil excavations is:

   I.   One with the ability to detect conditions which could result in cave-ins, failures in protective systems, hazardous atmospheres, and other hazards including those associated with confined spaces.
   II. One with the authority to take prompt corrective measures to eliminate existing and predicted hazards and to stop work when required.
   III. One with training, experience, and knowledge of soil analysis, use of protective systems, and the OSHA regulations pertaining to excavations.
   IV. One who is either a registered professional engineer or licensed geologist

   (A) II and III.
   (B) I, II, and III.
   (C) IV only.
   (D) All of the above.

END

CIVIL PE PRACTICE EXAM

BREADTH EXAM

VERSION C

SOLUTIONS

Problem	Solution	Problem	Solution
101	B	121	B
102	C	122	A
103	B	123	A
104	B	124	C
105	C	125	B
106	D	126	B
107	B	127	A
108	B	128	D
109	A	129	A
110	B	130	D
111	A	131	D
112	C	132	B
113	B	133	C
114	B	134	D
115	D	135	D
116	A	136	D
117	A	137	D
118	B	138	A
119	C	139	D
120	C	140	B

Solutions

101. Pipe Bedding QTO
Primary Category: Project Planning
Secondary Category: Quantity Take-Off Methods

Approach: Use the trench detail to calculate the volume of bedding required, then use the given density to calculate the weight of bedding required. To calculate the volume of bedding required, first calculate the cross-sectional area of the pipe:

$$A = \pi r^2 = \pi \cdot \left(\frac{12\ in}{2} \cdot \frac{1\ ft}{12\ in}\right)^2 = 0.785\ sf$$

Next calculate the rectangular area of the entire bedding area, including the pipe:

$$A = 3\ ft \cdot (12\ in + 2 * 6\ in) \cdot \frac{1\ ft}{12\ in} = 6\ sf$$

Now subtract the pipe area from the larger area to find the actual area of bedding material:

$$A = 6\ sf - 0.785\ sf = 5.215\ sf$$

Finally, calculate the weight with the allowance for waste:

$$W = 1.1 \cdot 300\ ft \cdot 5.215\ sf \cdot \frac{135\ lb}{cf} \cdot \frac{1\ ton}{2,000\ lb} = 116\ tons$$

102. **Cost Index**
Primary Category: Project Planning
Secondary Category: Cost Estimation

Approach: Use linear interpolation to calculate the cost indices for 2006 and 2018. Use the cost indices for year and location to adjust the cost of the old project to apply to the conditions for the new project.

$$2006 \; Cost \; Index: 1.00 + \frac{1.13 - 1.00}{2010 - 2005} \cdot (2006 - 2005) = 1.026$$

$$2018 \; Cost \; Index: 1.27 + \frac{1.47 - 1.27}{2020 - 2015} \cdot (2018 - 2015) = 1.39$$

$$Adjust \; Cost = \$1,200,000 \cdot \frac{1.39}{1.026} \cdot \frac{1.18}{0.98} = \$1,957,500$$

103. **Overtime**
     Primary Category: Project Planning
     Secondary Category: Cost Estimation

Approach: Calculate the number of hours required to complete the task at the supplied production rate. Next, calculate the excavator cost at the supplied straight rate and calculate the operator cost at straight time for the first 8 hours of every day and overtime for the last 2 hours of every day.

$$hrs\ req'd = \frac{165\ CY}{10\ CY/Hr} = 16.5 \quad hrs$$

$$Excavator\ cost = 16.5\ hrs \cdot \frac{\$100}{hr} = \$1,650$$

The first 10-hour day, the operator will work 8 hours of straight time and 2 hours of overtime. The second day the operator will only need to work 6.5 hours of straight time. There will be no overtime worked on the last day.

$$Operator\ cost = 2\ OT\ hrs \cdot \frac{1.5 \cdot \$50}{hr\ OT} + (8 + 6.5)\ ST\ hrs \cdot \frac{\$50}{hr\ ST} = \$875$$

$$Total = \$1,650 + \$875 = \$2,525$$

104.   **CPM Schedule**
        Primary Category: Project Planning
        Secondary Category: Project Schedules

        Approach: Identify the critical path. Activities on the critical path cannot be extended without increasing the total project duration. The critical path is A>B>E>F>H, since that path results in the longest duration, so only Activity B cannot be extended.

105.   **Rigging Load**
        Primary Category: Means and Methods
        Secondary Category: Construction Loads

        Approach: First calculate the load the rigging will support, then apply the factor of safety and convert to tons. Select the next largest size. Since the rigging is doubled around a pully, each rigging will support exactly half of the load:

$$Min. Rigging\ Strength = \frac{3,500\ lb}{2} \cdot \frac{1\ ton}{2,000\ lb} \cdot 5 \approx 4.4\ ton$$

        Of the options given, the 5-ton breaking strength is the minimum that is allowed.

106. **Utility Separation**
    Primary Category: Means and Methods
    Secondary Category: Construction Methods

    Approach: First calculate the maximum allowable invert elevation of the 8" diameter sewer at the crossing, then calculate the minimum required slope to achieve that invert elevation based on the upstream invert of the sewer and the distance to the crossing.

    $$Max\ Allowable\ IE\ @\ Crossing: 12.53\ ft - \frac{18\ in}{12\ ^{in}/_{ft}} - \frac{8\ in}{12\ ^{in}/_{ft}} = 10.36\ ft$$

    $$Min.\ Req'd\ Slope = \frac{12.30\ ft - 10.36\ ft}{50\ ft} \approx 3.9\%$$

107. **Steel Plate**
    Primary Category: Means and Methods
    Secondary Category: Temporary Structures and Facilities

    Approach: Since the steel plate is assumed to act like a 10-foot long simply supported beam, the maximum moment under a 5,000 pound point load can be calculated directly from the equation below:

    $$M_{max} = \frac{PL}{4} = \frac{5,000\ lb \cdot 1\ ^{kip}/_{1,000\ lb} \cdot 10\ ft}{4} = 12.5\ kip\ ft$$

108. **Active/Passive Earth Pressure**
     Primary Category: Soil Mechanics
     Secondary Category: Lateral Earth Pressure

Approach: Active earth pressures are mobilized behind a retaining wall, and passive earth pressures are mobilized in front of a retaining wall. Therefore, the active earth pressure resultant magnitude can be calculated as:

$$k_a = \left(\tan\left[45° - \frac{\emptyset}{2}\right]\right)^2 = \left(\tan\left[45° - \frac{32}{2}\right]\right)^2 \approx 0.31$$

$$R_A = \frac{1}{2}k_a\gamma H^2 = \frac{1}{2} \cdot 0.31 \cdot 120 \ pcf \cdot (10 \ ft)^2 \approx 1,860 \ lb$$

The passive earth pressure can be calculated as:

$$k_p = \left(\tan\left[45° + \frac{\emptyset}{2}\right]\right)^2 = \left(\tan\left[45° + \frac{32}{2}\right]\right)^2 \approx 3.25$$

$$R_P = \frac{1}{2}k_P\gamma H^2 = \frac{1}{2} \cdot 3.25 \cdot 120 \ pcf \cdot (5 \ ft)^2 \approx 4,875 \ lb$$

And the difference in magnitude is:

$$4,875 \ lb - 1,860 \ lb = 3,015 \ lb$$

109.   **Consolidation**
        Primary Category: Soil Mechanics
        Secondary Category: Soil Consolidation

Approach: An overconsolidated clay is one which has been exposed to a vertical pressure in the past (the preconsolidation pressure) which is higher than the current overburden pressure. A normally consolidated clay is one in which the current overburden pressure is equal to the highest vertical pressure the soil has ever experienced. The concept of preconsolidation pressure is not meaningful for a normally consolidated clay.

An overconsolidated clay's consolidation curve is steeper when overburden pressure exceeds the preconsolidation pressure. The slope of the consolidation curve above the preconsolidation curve on a log scale is called the compression index. The slope of the consolidation curve below the preconsolidation curve on a log scale is called the recompression index. The compression index is typically several times higher than the recompression index.

110. **Effective Stress**
     Primary Category: Soil Mechanics
     Secondary Category: Effective and Total Stresses

     Approach: Pay close attention to the soil unit weights that are provided. Recall that effective stress is the total stress minus the pore water pressure:

     $$\sigma' = \sigma - u$$

     Effective stress is that portion of the total stress that is supported through grain-to-grain contact. For the sand we need to calculate the in situ wet density (unit weight) of the soil using the given dry density both above and below the water table. We can use the following soil phase relationship:

     $$\gamma = \gamma_d(1 + w)$$

     For the clay, we can calculate the effective stress directly from the buoyant unit weight, because buoyant unit weight is defined as the saturated unit weight minus the unit weight of water:

     $$\gamma_b = \gamma_{sat} - \gamma_w$$

     Therefore:

     $$\gamma = \gamma_d(1 + w) = 105\ pcf \cdot (1 + 0.10) = 115.5\ pcf\ above\ GW$$

     $$\gamma = \gamma_d(1 + w) = 105\ pcf \cdot (1 + 0.35) = 141.75\ pcf\ below\ GW$$

     Therefore the effective stress is:

     $$115.5\ pcf \cdot 4\ ft + (141.75\ pcf - 62.4\ pcf) \cdot 2\ ft + 50\ pcf \cdot 5\ ft \approx 871\ psf$$

111. **Clay Bearing Capacity**
     Primary Category: Soil Mechanics
     Secondary Category: Bearing Capacity

     Approach: Bearing capacities for undrained, saturated clays are calculated assuming that the friction angle is equal to zero. For a friction angle equal to zero, the bearing capacity factors are:

     - $N_c = 5.14$
     - $N_q = 1.0$
     - $N_\gamma = 0.0$

     Recall also that cohesion is equal to one half of the unconfined compressive strength. There is no applied surface surcharge. Therefore, the bearing capacity equation is:

     $$q_{ult} = cN_c + \gamma D_f N_q + 0.5\gamma B_f N_\gamma$$

     $$q_{ult} = \frac{1500\ psf}{2} \cdot 5.14 + 110(2) = 4{,}075\ psf$$

     $$q_{all} = \frac{q_{ult}}{3} \approx 1{,}360\ psf$$

112.   **Consolidation Rate**
       Primary Category: Soil Mechanics
       Secondary Category: Foundation Settlement

Approach: The governing equation for primary consolidation rate is given below:

$$t = \frac{T_v H_d^2}{C_v}$$

Since we are told that the clay layer is 20 feet thick and has two-way drainage:

$$H_d = \frac{20\ ft}{2} = 10\ ft$$

Finally, the time factor T_v is obtained from a table published in a variety of references as a function of degree of consolidation. For 90%, $T_v = 0.848$. Therefore,

$$t = \frac{T_v H_d^2}{C_v} = \frac{0.848 \cdot (10\ ft)^2}{.0029\ sf/day} = 29{,}241\ days \approx 80\ years$$

113. **Undrained Slope Stability**
Primary Category: Soil Mechanics
Secondary Category: Slope Stability

Approach: The stability charts for undrained, cohesive soils require two inputs: slope angle in degrees and the dimensionless depth factor. The slope angle is:

$$\beta = \tan^{-1}\left(\frac{1}{2}\right) \approx 26.6°$$

The depth factor is the ratio of the distance to the underlying firm base to the depth of the cut:

$$d = \frac{D}{H} = \frac{5\ ft}{10\ ft} = 0.5$$

With these two values, the following stability number can be obtained from the chart:

$$N_0 \approx 6.3$$

Finally, the factor of safety for worst-case loading can be calculated as:

$$FS = \frac{N_0 c}{\gamma H} = \frac{6.3 \cdot 500\ psf}{120\ pcf \cdot 10\ ft} \approx 2.6$$

114. **Dead and Live Loads**
Primary Category: Structural Mechanics
Secondary Category: Dead and Live Loads

Approach: Calculate each load combination to determine which one results in the highest load.

I. $1.4(1,000\ psf) = 1,400\ psf$

II. $1.2(1,000\ psf) + 1.6(750\ psf) + 0.5(250\ psf) = 2,525\ psf$

III. $1.2(1,000\ psf) + (300\ psf) + (750\ psf) + 0.5(250\ psf) = 2,375\ psf$

IV. $1.2(1,000\ psf) + (500\ psf) + (750\ psf) + 0.2(250\ psf) = 2,500\ psf$

115. **Trusses**
Primary Category: Structural Mechanics
Secondary Category: Trusses

Approach: Begin by calculating the reaction forces by summing the forces in the y-direction and taking the moment about one reaction:

$$\sum F_y = 0 = R_L + R_R - 10 \ kip - 5 \ kip; \ \therefore \ R_L + R_R = 15 \ kip$$

$$\sum M_L = 0 = -10 \ kip \cdot 15 \ ft - 5 \ kip \cdot 30 \ ft + R_R \cdot 45 \ ft; \ \therefore \ R_R \approx 6.7 \ kip; \ \therefore \ R_L \approx 8.3 \ kip$$

Next, slice the truss as shown:

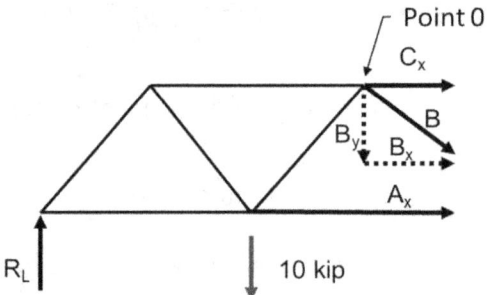

Take the moment about Point "0" and solve for the force in member A, which is only in the x-direction.

$$\sum M_0 = 0 = -8.3 \ kip \cdot \left(15 + \frac{15}{2}\right) ft + 10 \ kip \cdot \left(\frac{15}{2}\right) ft + A \cdot 10 \ ft; \ \therefore \ A = -11.2 \ kip$$

116. **Bending Stress**
     Primary Category: Structural Mechanics
     Secondary Category: Bending Stress

Approach: Bending stress in a beam is distributed linearly through the cross section. At the midpoint, or neutral axis, the bending stress is zero. It increases linearly as distance from the neutral axis increases, with half of the section in tension and half the section in compression. The magnitude of the bending stress at any distance from the neutral axis can be calculated as:

$$\sigma_b = \frac{My}{I_c}$$

Where M is the moment, y is the distance from the neutral axis, and I_c is the moment of inertia about the centroid of the cross section. The required parameters can be calculated as follows:

$$M = \frac{PL}{4} = \frac{5 \; kip \cdot 10 \; ft \cdot \dfrac{12 \; in}{ft}}{4} = 150 \; kip \; in$$

$$I_c = \frac{bh^3}{12} = \frac{4 \; in \cdot (8 \; in)^3}{12} \approx 170.7 \; in^4$$

Since the neutral axis is at mid-depth, Point "A" is 2 inches from the neutral axis:

$$y = 2 \; in$$

$$\sigma_b = \frac{(150 \; kip \; in)(2 \; in)}{170.7 \; in^4} \approx 1.8 \; ksi$$

117. **Shear Diagrams**
Primary Category: Structural Mechanics
Secondary Category: Shear Forces

Approach: Shear diagrams for this uniformly loaded beam must comply with the following rules:
- The change in shear at the supports should be equal to the reaction support force. In this case both support reactions should be positive, and intuition tells us the right support force should be larger than the left support force.
- The magnitude of the uniform load is equal to the slope of the shear.
- The shear diagram must resolve to zero at each of the supports.

Only one shear diagram complies with these rules.

118. **Deflection**
Primary Category: Structural Mechanics
Secondary Category: Deflection

Approach: Maximum deflection for a cantilever beam with end load can be calculated directly as:

$$y_{max} = \frac{PL^3}{3EI} = \frac{5\ kip \cdot 1000\frac{lb}{kip} \cdot \left(10\ ft \cdot 12\ ^{in}/_{ft}\right)^3}{3 \cdot 29 \cdot 10^6 \cdot I} = \frac{99.31}{I}$$

Using this formula, the deflection for each shape can be calculated:

Shape	Z_x (in^3)	I_x (in^4)	E (10^6 psi)	y_{max} (in)
W14x86	126	1140	29	0.09
W14x74	126	795	29	0.12
W12x79	119	662	29	0.15
W10x88	113	534	29	0.19

Select the most economical (lightest) option that limits deflection to the prescribed maximum.

119. **Axial Force**
     Primary Category: Structural Mechanics
     Secondary Category: Axial stress/forces

     Approach: The modulus of elasticity can be used directly to relate axial stress to strain:

$$E = \frac{stress}{strain} = \frac{P/A}{\Delta L/L} = 30,000\ ksi = \frac{P/\left(\pi\left(2\ in \cdot 1/2\right)^2\right)}{0.5\ in \bigg/ \left(100\ ft \cdot 12\ in/ft\right)}; \quad \therefore P \approx 40\ kip$$

120. **Open Channel**
     Primary Category: Hydraulics and Hydrology
     Secondary Category: Open Channel Flow

     Approach: Use the Manning Equation to calculate the flowrate in the channel. Then, calculate the volume of the pond at the stipulated depth. Finally, divide the volume by the flowrate to find the required time.

$$A = \frac{1}{2}bh = \frac{1}{2} \cdot \frac{18\ in}{12\ in/ft} \cdot \frac{12\ in}{12\ in/ft} = 0.75\ sf$$

$$P_w = 2 \cdot \sqrt{\left(\frac{9\ in}{12\ in/ft}\right)^2 + \left(\frac{12\ in}{12\ in/ft}\right)^2} = 2.5\ ft$$

$$Q = \frac{1.49}{n}AR^{2/3}S^{1/2} = \frac{1.49}{0.02} \cdot 0.75\ sf \cdot \left(\frac{0.75\ sf}{2.5\ ft}\right)^{\frac{2}{3}} \cdot 0.01^{\frac{1}{2}} \approx 2.5\ cfs$$

$$t = \frac{V}{Q} = \frac{1,000\ sf \cdot 3\ ft}{2.5\ cfs} \cdot \frac{1\ min}{60\ sec} \approx 20\ min$$

121. **Stormwater Pipes**
     Primary Category: Hydraulics and Hydrology
     Secondary Category: Stormwater collection and drainage

     Approach: First determine the required overflow rate by subtracting the treatment rate from the peak discharge. Then calculate the required pipe diameter using the Manning equation for a pipe flowing full.

$$Q = 5 \, cfs - \frac{1000 \, gpm}{448.8 \, ^{gpm}/_{cfs}} = 2.77 \, cfs$$

$$D = \left[\frac{C_0 Qn}{\sqrt{S}}\right]^{3/8} = \left[\frac{2.16(2.77)(0.013)}{\sqrt{.02}}\right]^{3/8} = \approx 0.8 \, ft \approx 9.6 \, in$$

122. **Rainfall Distributions**
     Primary Category: Hydraulics and Hydrology
     Secondary Category: Storm characteristics

     Approach: The SCS Type 2 storm has a greater peak discharge than the SCS Type 1A storm, as evident from inspection of the steepness of the curves and as widely reported in the literature. Both storms result in the same total precipitation and same total runoff volume, because they are unit hydrographs. The SCS Type 1A peak discharge rate occurs earlier than the SCS Type 2 peak discharge rate, as evident from inspection of the graph and as widely reported in the literature.

123. **Rainfall Distributions**
    Primary Category: Hydraulics and Hydrology
    Secondary Category: Runoff Analysis

    Approach: First determine the representative runoff coefficient for the entire basin by weighting each type of land cover by area:

    $$C = \frac{1}{25\ ac} \cdot (10\ ac \cdot 0.4 + 4\ ac \cdot 0.75 + 6\ ac \cdot 0.85 + 5\ ac \cdot 0.9) = 0.664$$

    Next calculate the rainfall intensity with the given time of concentration.

    $$I(^{in}/_{hr}) = \frac{200}{time\ of\ concentration\ (min)} = \frac{200}{60} \approx 3.333$$

    Finally, use the Rational Method to calculate the maximum runoff rate:

    $$Q = CIA = 0.664 \cdot 3.333 \cdot 25\ ac \approx 55\ cfs$$

124. **Pond**
    Primary Category: Hydraulics and Hydrology
    Secondary Category: Detention/retention ponds

    Approach: First calculate the net flowrate into the pond:

    $$Q_{in} = 500\ gpm \cdot \frac{1\ cfs}{448.8\ gpm} \approx 1.11\ cfs$$

    $$Q_{out} = 50\ ft \cdot 50\ ft \cdot \frac{10\ in}{hr} \cdot \frac{1\ ft}{12\ in} \cdot \frac{1\ hr}{60\ min} \cdot \frac{1\ min}{60\ sec} \approx 0.58\ cfs$$

    $$Q = 1.11\ cfs - 0.58\ cfs = 0.53\ cfs$$

    In one hour the total volume into the pond is:

    $$V = Qt = 0.53\ cfs \cdot 1\ hr \cdot \frac{60\ min}{hr} \cdot \frac{60\ sec}{min} = 1{,}908\ cf$$

    The height required to achieve the volume is:

    $$V = 1{,}908\ cf = h \cdot 50\ ft \cdot 50\ ft; \quad \therefore h = 0.76\ ft \approx 9\ in$$

125. **Headloss**
Primary Category: Hydraulics and Hydrology
Secondary Category: Pressure conduit

Approach: Calculate the headloss in the pipe with the Darcy-Weisbach equation and the Hazen-Williams equation. All required variables are given. First Darcy-Weisbach:

$$h_l = \frac{fl}{D} \cdot \frac{v^2}{2g}$$

$$h_l = \frac{0.018 \cdot 100\ ft}{\dfrac{8\ in}{12\ in/ft}} \cdot \frac{\left(\dfrac{\left(\dfrac{3000\ gpm}{448.8\frac{gpm}{cfs}}\right)}{\left(\pi \cdot \left(\dfrac{8\ in}{2 \cdot 12\frac{in}{ft}}\right)^2\right)}\right)^2}{2 \cdot 32.2\ \frac{ft}{s^2}} \approx 15.4\ ft$$

Then Hazen-Williams:

$$h_{l,ft} = \frac{10.44 L_{ft} Q_{gpm}^{1.85}}{C^{1.85} d_{in}^{4.87}}$$

$$h_{l,ft} = \frac{10.44 \cdot 100\ ft \cdot (3{,}000\ gpm)^{1.85}}{120^{1.85} \cdot (8\ in)^{4.87}} \approx 16.1\ ft$$

126. **Energy Grade Lines**
Primary Category: Hydraulics and Hydrology
Secondary Category: Pressure conduit

Approach: The top line is the energy grade line and the bottom line is the hydraulic grade line. The difference between the two (Dimension C) is the velocity head. Dimension B is the total loss in energy over the piping system, the drop in the energy grade line.

127.   **Horizontal Curves**
        Primary Category: Geometrics
        Secondary Category: Basic circular curve elements

        Approach: The following horizontal curve equations can be used to solve for the degree of curvature:

$$C = 2R \sin\frac{I}{2}; \quad 850\ ft = 2R \sin\frac{25°}{2}; \quad \therefore R \approx 1963.6\ ft$$

$$R = 1963.6\ ft = \frac{5729.578}{D}; \quad \therefore D \approx 2.9°$$

128.   **Vertical Curves**
        Primary Category: Geometrics
        Secondary Category: Basic vertical curve elements

        Approach: The k-value is defined as the length of the curve over the change in grade:

$$k = \frac{L}{|G_2 - G_1|}$$

Therefore:

$$120 = \frac{L}{|-0.5 - 2.5|}; \quad \therefore L = 360$$

The curve must be at least 360 feet long. Select the next largest number of the options given.

129. **Peak Hour Factor**
Primary Category: Geometrics
Secondary Category: Traffic Volume

Approach: The peak hour factor is defined as the peak hourly volume divided by the peak rate of flow within that hour. Since the data was collected in 10-minute increments, the peak 10-minute flow must be multiplied by six (six ten-minute intervals per hour):

$$PHF = \frac{Peak\ Hourly\ Volume}{6\ \frac{periods}{hour} \cdot Peak\ Period\ Volume\ in\ that\ Hour}$$

From inspection of the data, the hourly volumes are:

Hour	Count
5:00-6:00	920
5:10-6:10	915
5:20-6:20	893
5:30-6:30	896
5:40-6:40	893
5:50-6:50	871
6:00-7:00	858

The peak hourly volume is 920 vehicles between 5:00 and 6:00. The peak 10-minute volume in that period is 162 vehicles (between 5:40 and 5:50). Therefore:

$$PHF = \frac{920\ vph}{6\ ^{periods}/_{hr} \cdot 162\ ^{veh}/_{period}} \approx 0.95$$

130. **Soil Classification**
Primary Category: Materials
Secondary Category: Soil Classification

Approach: The distinction between a CL and CH soil is made based on whether the liquid limit is greater than 50. Close examination of the plasticity chart shows that plasticity index and therefore plastic limit (the y-axis on the plasticity chart) separates clays from silts (i.e., CH vs. MH or CL vs. ML).

48

131. **CBR**
Primary Category: Materials
Secondary Category: Soil Properties

Approach: The California Bearing Ratio test (ASTM D1883) is used to determine the suitability of a soil for use as a pavement subgrade.

132. **Cement Types**
Primary Category: Materials
Secondary Category: Concrete

Approach: Highly alkaline soils require sulfate-resistant Portland cement (Type V cement). Additionally, a concrete mix typically has at least coarse aggregate, sand, cement, and water.

133. **Steel Yield Stress**
Primary Category: Materials
Secondary Category: Steel

Approach: The yield stress is the tensile stress at which the stress-strain curve first exhibits an increase in strain without a corresponding increase in stress. The shape of the curve shown is characteristic of most steels. The first maxima at 31 ksi is the yield stress by definition.

134. **Compaction**
    Primary Category: Materials
    Secondary Category: Soil Compaction

    Approach: Wet density and dry density can be related as follows:

    $$\gamma_d = \frac{\gamma}{1 + w}$$

    Therefore the required wet density at optimum moisture content and 95% of the maximum dry density as determined by the modified proctor is:

    $$0.95 \cdot 135 \, pcf = \frac{\gamma}{1 + 0.08}; \quad \therefore \gamma = 139 \, pcf$$

135. **Specification Conformance**
    Primary Category: Materials
    Secondary Category: Specification Conformance

    Approach: The calculated compressive strengths are shown below:

Test No.	Cylinder Diameter (in)	Axial Compressive Failure Load (lb)	Compressive Strength (psi)
1	6	91,584	3,239.12
2	6	92,574	3,274.14
3	6	70,548	2,495.13
		Average	3,002.79

Therefore the specification is not met because Test No. 3 is less than 2,500 psi.

136. **Shrink/Swell**
    Primary Category: Site Development
    Secondary Category: Excavation and Embankment

    Approach: Compacted cubic yards can be converted to loose cubic yards with the following equations, where SF is the swell factor, DF is the shrink factor, LCY are loose cubic yards, BCY are bank cubic yards, and CCY are compacted cubic yards.

    $$SF = \frac{LCY}{BCY}$$

    $$DF = 1 - \frac{CCY}{BCY}$$

    Therefore:

    $$0.15 = 1 - \frac{100\ CCY}{BCY}; \quad \therefore BCY \approx 117.6\ BCY$$

    $$1.25 = \frac{LCY}{117.6\ BCY}; \quad \therefore LCY \approx 147\ LCY$$

137. **Staking**
    Primary Category: Site Development
    Secondary Category: Construction Site Layout and Control

    Approach: Stake 1 indicates a fill of 3.1 feet is required. Stake 2 indicates a cut of 1.8 feet is required. Therefore the finished grade elevations are:

    $$Point\ 1\ FG\ Elev. = 81.5\ ft + 3.1\ ft = 84.6\ ft$$

    $$Point\ 2\ FG\ Elev. = 79.7\ ft - 1.8\ ft = 77.9\ ft$$

    And the difference is:

    $$84.6\ ft - 77.9\ ft = 6.7\ ft$$

138. **TESC**
Primary Category: Site Development
Secondary Category: Temporary and permanent erosion and sediment control

Approach: The primary purpose of erosion and sediment control provisions is to protect water quality and prevent streambank erosion.

139. **OSHA 1926**
Primary Category: Site Development
Secondary Category: Safety

Approach: Section 1926 of the Code of Federal Regulations contains the safety regulations that pertain specifically to the construction industry.

140. **Competent Person**
Primary Category: Site Development
Secondary Category: Safety

Approach: There is no OSHA requirement for a competent person to hold an engineering or geology license. All the other requirements apply.